THE WORLD'S TRANSPORT

ROAD TRAVEL

Tim Wood

Wayland

THE WORLD'S TRANSPORT

Air Travel

Rail Travel

Road Travel

Water Travel

All words that appear in **bold** are explained in the glossary.

Editor: James Kerr
Designer: Loraine Hayes

Cover: A freight truck in Nevada, USA.

First published in 1992 by
Wayland Publishers Ltd
61 Western Road, Hove
East Sussex BN3 1JD, England

British Library Cataloguing in Publication Data

Wood, Tim
    Road travel.—(The world's transport)
    I. Title   II. Series
    388.309

ISBN 0-7502-0373-0

Typeset by Dorchester Typesetting Group Ltd
Printed in Italy by G. Canale & C.S.p.A.
Bound in France by A.G.M.

CONTENTS

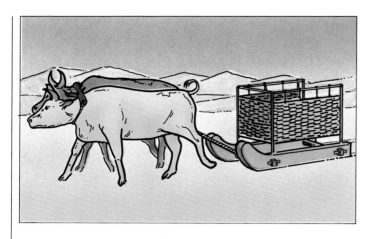

A Sumerian cart.

A Roman chariot.

THE DEVELOPMENT OF ROAD TRAVEL

The first travellers were Stone Age hunters searching for game. Their 'roads' were the trails made by wild animals.

The wheel, which was invented in Sumeria in about 3,000 BC, revolutionized transport. Soon after, the cart and harness was invented. As trade between towns and cities developed, people began to build roads. The first great road engineers were the Romans.

They knew how to lay strong foundations and make a smooth surface with flat stones.

After the collapse of the Roman Empire in about 420 AD, roads in Europe fell into disrepair. It was not until around 1700, with the start of the **Industrial Revolution**, that roads and transport began to improve. At first, travel was by stagecoach and cart. Then, in 1885, the German Gottlieb Daimler

An eighteenth-century stagecoach.

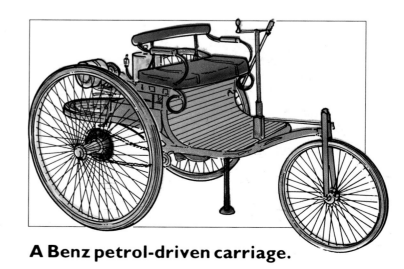

A Benz petrol-driven carriage.

A Ford Model T.

Mass-production of Model Ts.

built the first engine powered by petrol. He worked with Karl Benz to make the first petrol-driven carriage. The age of the motor car had arrived.

Throughout the 1920s, the motor industry grew rapidly. The main reason for this was the introduction of the **assembly line** in 1913 by Henry Ford. This meant that the Ford Model T could be assembled in an hour and a half — about eight times faster than the old method of hand-building.

Mass-production of vehicles led to a huge increase in the number of motor cars on the roads, and a growing number of road accidents. Improvements, such as tarred road surfaces, traffic lights and pedestrian crossings, were introduced in an attempt to improve road safety. Travel became more convenient as petrol stations were built.

Motor car design slowly improved as well. Many features of modern cars were introduced during the 1920s and 1930s. These included **hydraulic** four-wheel brakes, safety glass, bumpers, **pneumatic tyres**, electric lights and horns, windscreen wipers, **speedometers** and car radios.

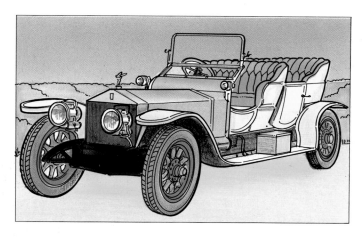
A Rolls-Royce Silver Ghost.

DID YOU KNOW?

- **Roads are so old that no one is sure where the word 'road' comes from. It may have come from the Old English word 'rad' meaning to ride. Britain's oldest road was built in 4,000 BC.**
- **The Romans built over 80,000 km of roads. Some of them are still in use today.**
- **In Ancient Assyria, anyone who parked a chariot or cart in the wrong place in the capital city of Nineveh could be put to death by being impaled on a stake.**

MOVING ON TWO WHEELS

The first bicycle, invented in about 1790, was wooden and had no pedals. The rider sat on the crossbar and 'scooted' the bicycle along with his or her feet. In about 1839, a Scottish blacksmith, Kirkpatrick Macmillan, made the first cycle with pedals. The success of this machine led to the invention of many different types of bicycles. One of the most famous early bicycles was the 'penny-farthing', which

ABOVE The mountain bike is now used to travel round towns and cities.
BELOW A modern racing bicycle.

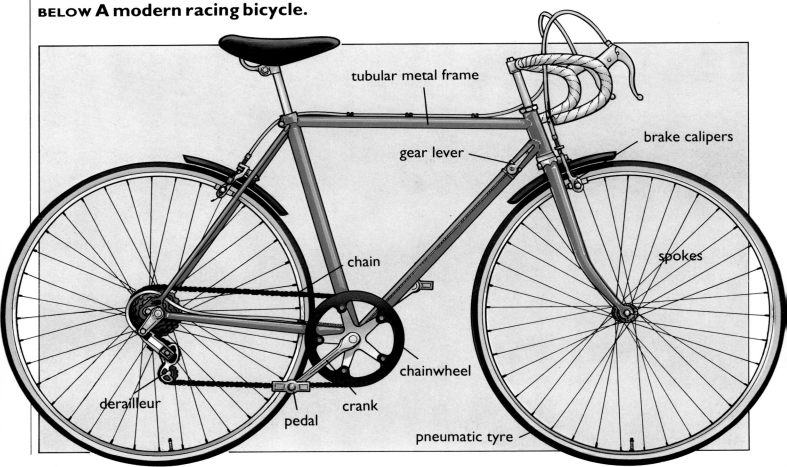

tubular metal frame

gear lever

brake calipers

spokes

chain

chainwheel

derailleur

crank

pedal

pneumatic tyre

There are over 210 million bicycles and tricycles in China.

had a huge front wheel up to 2.2 m wide. Each complete rotation of the pedals turned the wheel round once, so these cycles were only suitable for very athletic riders!

By about 1890, 'safety' bicycles with metal frames, pedals, chains, pneumatic tyres and brakes were a common sight on the roads.

Cycling has increased in popularity in the last twenty years. This is partly because many people have become more concerned about their health and physical fitness, and cycling is very good exercise. Also, many people claim that cycles are quicker and more convenient than motor cars for going to work in today's overcrowded cities. Since it causes no pollution, the bicycle is the most 'environmentally friendly' form of road transport.

> **DID YOU KNOW?**
> - **Kirkpatrick Macmillan committed the first cycling offence when he knocked over a child in Glasgow in 1842. He was fined five shillings (25p).**

FOUR IMPORTANT INVENTIONS

There are four important developments which have made road travel what it is today.

The internal combustion engine
This engine, which produces power by burning fuel inside itself, is the main part of a motor car. The first time an internal combustion engine was used to power a vehicle was in 1826 when an Englishman, Samuel Brown, built a gas-driven carriage. The first internal combustion engine to run on petroleum was invented by Daimler in 1885.

Tarmac roads
Before 1850, roads were only suitable for horse-drawn traffic. As road traffic increased towards the end of the 1900s, it became necessary to build better, longer-lasting roads with smooth, waterproof surfaces. Tarmac, first used in road construction in 1845, greatly improved the quality of roads.

LEFT **Most modern roads have strong foundations made of concrete and stone. On top of this are layers of tarmac – a mixture of crushed stone for strength, and tar to stick the stones together and provide a waterproof seal. Most roads have a slightly curved surface so that rainwater runs away from the centre into drains at the sides.**

Road

tarmac

crushed stone   concrete base

Motorway

rough concrete surface

crushed stone   concrete   steel mesh

The electrical system

Early cars had very simple electrical systems which worked only the **ignition**. The first electric self-starter was fitted to an Arnold Sociable in 1896 and the first electric lamps for cars were produced in 1908. In 1912, a Cadillac became the first car to go on sale with a self-contained electrical system for ignition, starting and lighting.

Tyres

Vehicles were originally fitted with solid rubber tyres which made the ride very bumpy. This changed in 1895, when Edouard Michelin became the first person to fit pneumatic tyres to a car. Pneumatic tyres are hollow rubber tubes which are filled with compressed air. This absorbs the shock as the vehicle passes over a bump.

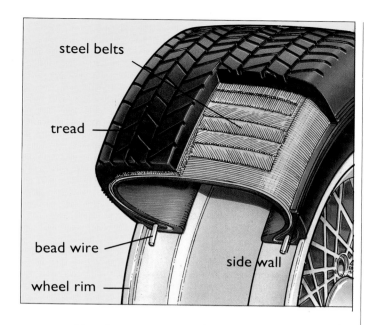

ABOVE **Early pneumatic tyres had rubber inner tubes. Most modern tyres are tubeless and are made of rubber reinforced with fabric and steel belts. The air pressure inside the tyre holds it firmly on to the wheel rim.**

Motorways need to be stronger than ordinary roads. They have concrete bases which are strengthened with steel.

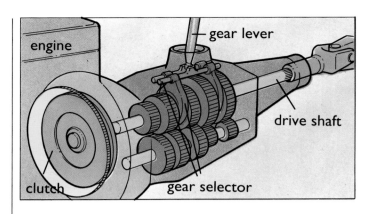

engine

gear lever

drive shaft

gear selector

clutch

A gear is a toothed cog-wheel attached to a turning shaft or rod. Gears are used to pass on movement from one turning shaft to another and change the speed at which the shaft turns. Manual gearboxes (above) contain a number of toothed cog-wheels of different sizes. As the driver changes from a lower gear to a higher one, the amount of turning power supplied to the wheels increases and the car can go faster.

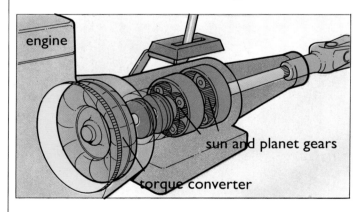

engine

sun and planet gears

torque converter

In a manual gearbox, the driver changes gear by moving a gear lever. The clutch disconnects the engine from the gearbox while the gears are being changed. Some cars have automatic gearboxes, like the one above, which change gear as the speed of the engine changes. The driver does not use a gear lever.

HOW THE MOTOR CAR WORKS

Most motor cars are powered by petrol engines. When the driver turns the ignition key, an electrical current from the battery provides power to turn the starter motor and pump petrol from the fuel tank to the carburettor. The petrol is mixed with air in the carburettor. As the starter motor turns the engine, the driver presses the accelerator pedal. This draws the mixture of petrol and air into cylinders inside the engine. Each cylinder has a piston inside it. The piston moves up and squashes the mixture. A spark then sets the mixture alight, producing an explosion. This explosion forces the piston downwards, which turns the crankshaft. The crankshaft is a special rod that runs through the engine. As the crankshaft turns, it pushes the piston up. The mixture is squashed again and the process is repeated.

The pistons move up and down with a quick pumping action, turning the crankshaft. The crankshaft carries the power produced by the engine, through the **transmission system**, to the wheels. The turning power the engine delivers to the wheels is controlled by **gears** contained in the gearbox.

Waste gases are released through the exhaust pipe. The engine is cooled by water which passes round a metal jacket surrounding the engine. Some engines are cooled by a powerful fan.

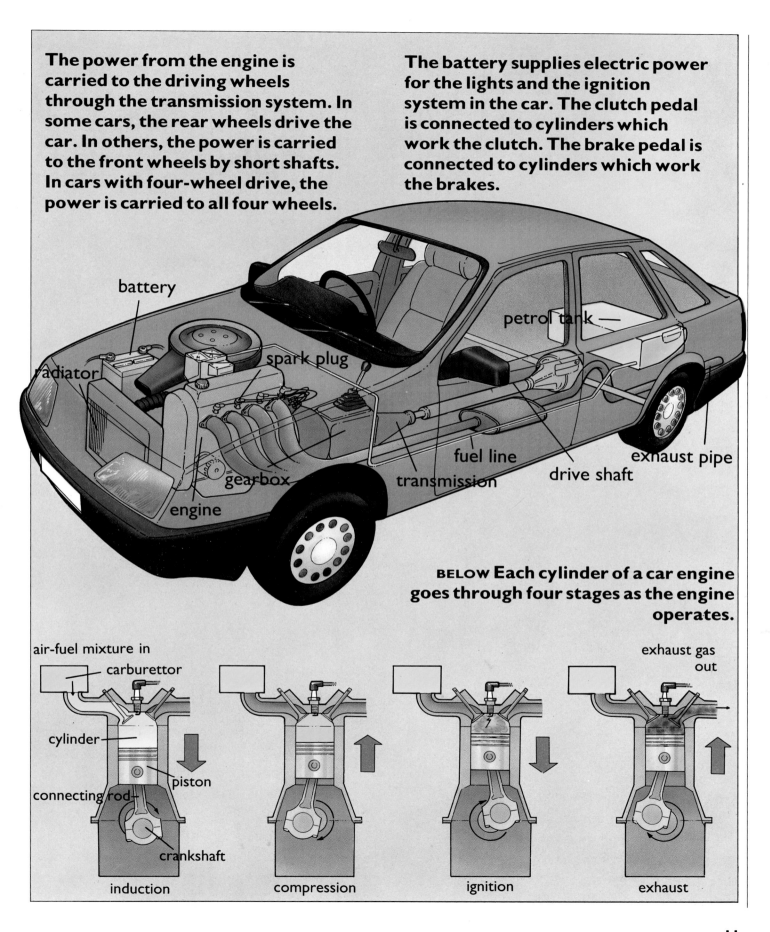

The power from the engine is carried to the driving wheels through the transmission system. In some cars, the rear wheels drive the car. In others, the power is carried to the front wheels by short shafts. In cars with four-wheel drive, the power is carried to all four wheels.

The battery supplies electric power for the lights and the ignition system in the car. The clutch pedal is connected to cylinders which work the clutch. The brake pedal is connected to cylinders which work the brakes.

battery

petrol tank

radiator

spark plug

fuel line

exhaust pipe

gearbox

drive shaft

transmission

engine

BELOW Each cylinder of a car engine goes through four stages as the engine operates.

air-fuel mixture in

exhaust gas out

carburettor

cylinder

piston

connecting rod

crankshaft

induction

compression

ignition

exhaust

CAR PRODUCTION

Designing and producing a new motor car requires several years of research and development. The process begins with sketches made by designers. These sketches are then turned into clear design drawings with the aid of a computer. A computer-aided design (CAD) system allows engineers to create a detailed electronic image of the car on the computer screen. This image can be rotated, enlarged and viewed from many different angles. From this engineers can produce complete details of the car. These will help the engineers to produce the parts of the car.

Clay, wood and plastic models of the car are also made from the design drawings. Some of these are **scale models**, others are full-sized. The models are used for a huge range of scientific experiments, such as wind tunnel tests which show the **aerodynamic** qualities of the car. When all the tests have been completed, a **prototype** is made. This is **scanned** electronically to produce exact information about the size of the car. The information is fed into the machines that press the body parts, so that they can make exact copies of each part of the car.

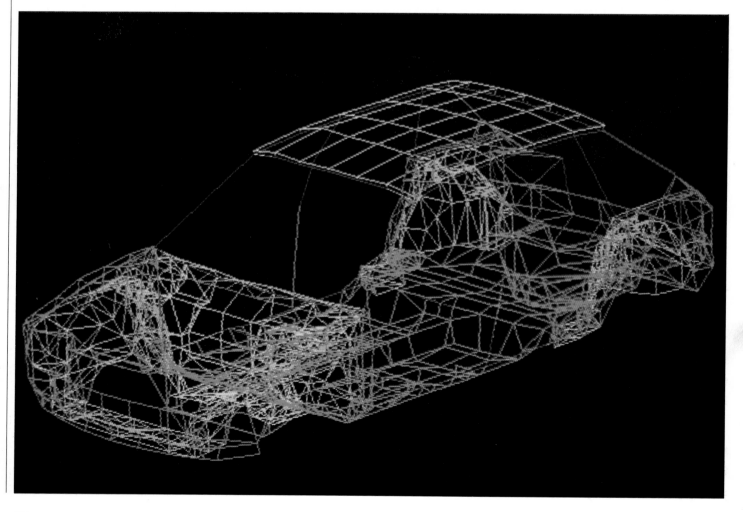

ABOVE **Although some cars are still
assembled by human workers, much of
the work in a modern car factory is
done by computer-controlled robot
machines.**

LEFT **Designers can use a CAD system
to run tests on a car, using computer
programmes.**

DID YOU KNOW?
- **Assembling a modern car
  involves nearly 2,000 separate
  operations and takes about two
  days from start to finish.**
- **There are 170 million motor
  vehicles in the USA.**

TRUCKS

Most trucks are powered by diesel engines. The truck starts when the air in the cylinder is squashed by the piston moving upwards. This makes the temperature of the air in the cylinder rise dramatically. When the piston rises as high as it can, fuel is injected into the cylinder as a fine spray. The heat of the air in the cylinder ignites the fuel—air mixture. The gases produced by the explosion drive the piston downwards. The power of some diesel engines is increased by a **turbocharger**.

Trucks usually have many more gears than cars. This allows them to travel in

Trucks in China transport goods to areas which cannot be reached by rail.

all sorts of different conditions. Since trucks weigh several tonnes when fully loaded, they need excellent brakes. Many trucks have powerful brakes worked by compressed air, instead of the hydraulic system used in cars. The engine supplies the power for the air compressor.

Many trucks have tippers, crushers, hoists or other special machinery. A pump driven by the engine pushes hydraulic fluid through a network of pipes and cylinders to work these machines. Most modern heavy trucks have **power-steering** mechanisms. Hydraulic pressure makes the power-steering mechanism work.

The engine of a big truck produces more than 400 horsepower (hp). A very powerful car produces about 200 hp.

BELOW **The main parts of a modern truck.**

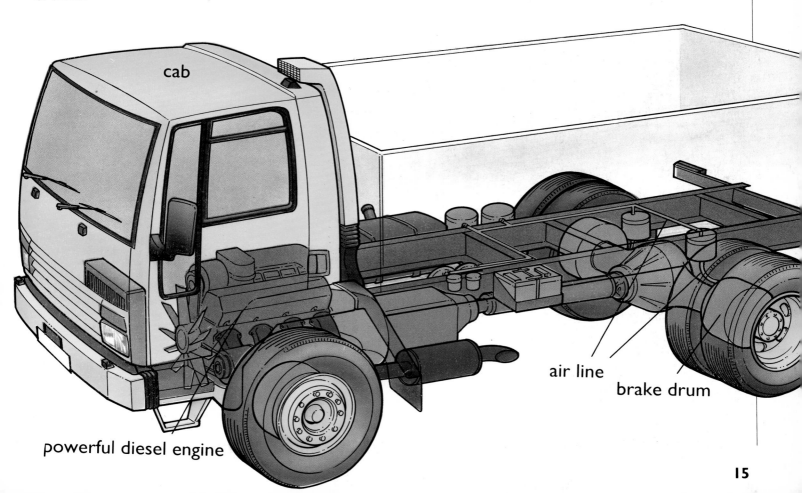

cab

air line

brake drum

powerful diesel engine

COMPUTER CONTROL

Computers play an important role in modern motoring. Many cars now have electronic equipment that makes them safer and easier to drive. Modern cars have few instruments that actually give the driver information. Instead they have a series of electronic warning-lights. These are meant to attract the driver's attention when there is a problem with a particular part of the car. For example, all modern cars have an oil warning-light which comes on when the engine needs more oil or the oil pressure drops. Many new cars have warning buzzers which sound if the lights are left on when the car is parked, if safety belts are not fastened, or if doors are not closed properly.

Some luxury cars have advanced computer systems. These work out and display information about the car's performance, including how much fuel is being used, and the average speed over the journey. Some on-board computers

On-board computer systems in luxury cars such as the Merlin 2000 can be programmed to give the driver the information he or she wants to know.

A computer terminal in a police car.

will work out the best route for the driver to take. Computerized engine-management systems make constant changes to the fuel and ignition system of the car. This makes sure that the car works as well as possible. Some cars are fitted with electronic terminals so that mechanics can connect the engine to a computer. The computer can find tiny faults, which allows the mechanic to adjust the engine so it can work in the best way.

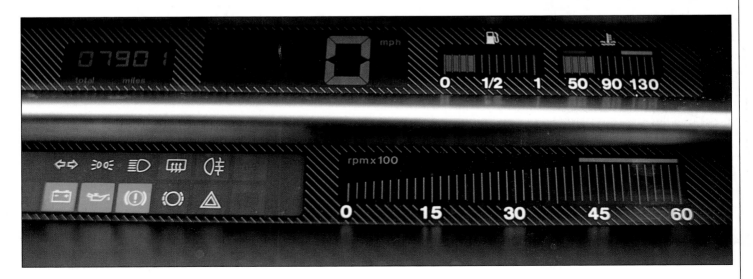

The panel in the bottom left-hand corner of this dashboard contains a series of warning lights. The middle red button is the oil warning-light.

MOTORBIKES

Early motorbikes were basically bicycles with engines. The first motorbike was invented by Gottlieb Daimler in 1885 and was mainly made of wood. However, Daimler only wanted it as a **test-bed** for his engine.

One of the problems for early motorbike inventors was to decide where to put the engine and how to transmit power to the driving wheel. The first really popular motorbike was developed by Heinrich Hildebrand and Alois Wolfmüller who built a two-cylinder bike in 1893. Other machines included one built by a Frenchman, Millet, which had a **radial engine** built into the back wheel. The motorbike built by the Werner brothers from Russia had the engine mounted over the front wheel, which was driven by a belt. By about 1900, motorbike designers had settled on something like the modern machine, with the engine mounted in a solid triangular frame between the two wheels.

Motorbikes are fast and can move easily along busy roads in traffic conditions that would delay a car or truck. For this reason, motorbikes are used by police throughout the world for chasing criminals and for traffic duties. In some countries, motorbikes are used to carry doctors or paramedics to the scenes of accidents. In many large cities, motorbikes are used by dispatch riders for the speedy delivery of packages and papers.

Modern Japanese superbikes, like the Honda VFR 750, are some of the most powerful machines on the road.

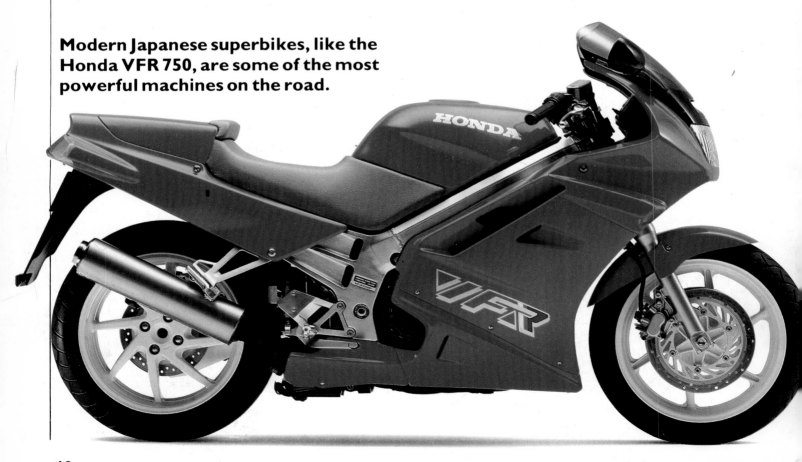

ABOVE **A Harley Davidson touring bike.**

BELOW **The main parts of a motorbike.**

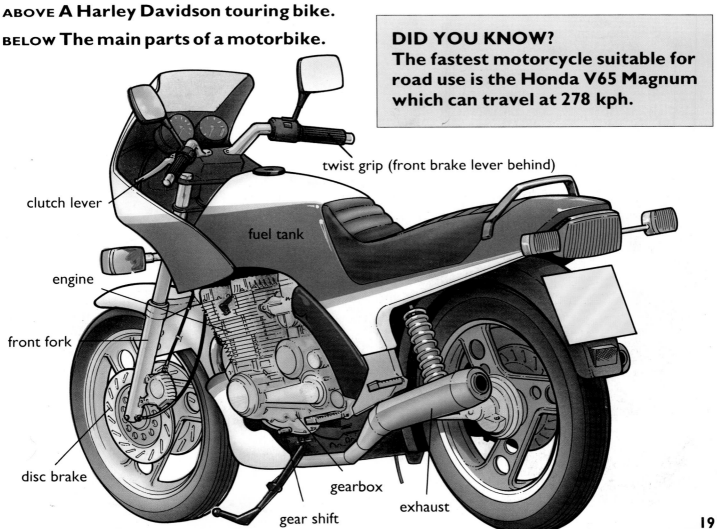

twist grip (front brake lever behind)

clutch lever

fuel tank

engine

front fork

disc brake

gearbox

gear shift

exhaust

DREAM MACHINES

In these days of mass-production, many people try to make their cars stand out from the crowd. Advertisers try to convince motorists that the rather ordinary family saloon cars most of them buy are the cars of their dreams. A few car and motorbike manufacturers actually do produce exotic 'dream' machines. Perhaps the ultimate **production car** is the 325 kph Lamborghini Diabolo. You can buy one of these if you have £155,000 to spend!

Those who cannot afford these dream machines, but still want to look different, customize their cars. They make them different, more streamlined, and more exciting by altering their shape and colour. A custom car is the ultimate form of personal transport because each one is unique and reflects the taste and imagination of the owner.

Custom cars come in all shapes and sizes. 'Street machines' have huge tyres and powerful engines. 'Street rods' are designed for illegal straight-line road races. This form of custom car has now developed into the full-blown drag racer. 'Low riders' are long, sleek, ground-hugging vehicles designed for cruising smoothly around the streets. All types of vehicle, including vans and pick-up trucks, have been customized.

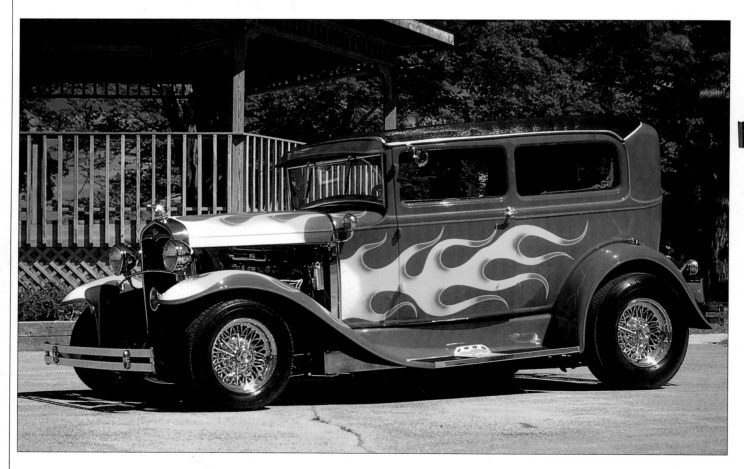

Spectacular paintwork is an essential part of a custom car.

DID YOU KNOW?
- The fastest steam car is *Steamin' Demon*, which reached 234 kph on the Bonneville Salt Flats, Utah, USA in 1985.
- The fastest car is *Thrust 2*, which was timed at 1019.4 kph on the Black Rock Desert, Nevada, USA in 1983.

The Lamborghini Countach is one of the fastest and sleekest cars suitable for road use.

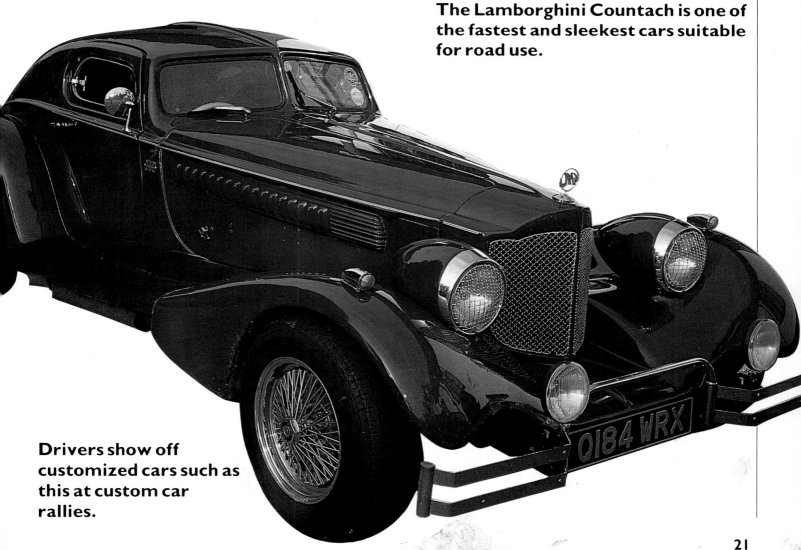

Drivers show off customized cars such as this at custom car rallies.

HOW MOTOR TRANSPORT CHANGED THE WORLD

Motor transport has had a great effect on our lives. The production, selling, driving and maintenance of cars is the world's biggest business, employing over 25 million people. The huge increase in motor vehicles has caused a massive growth in the oil industry, which has created new jobs and brought great wealth to some countries.

Since 1945 more and more roads have been built. Out-of-town shopping centres have sprung up in places which are convenient for people to reach by car. The motor car has become an essential part of life. In the USA, 'drive-in' shops, restaurants, banks, cinemas and even churches allow drivers to lead their lives without ever having to leave their vehicles!

Big changes have been necessary to keep the increasing amount of traffic

ABOVE **Taking the 'car culture' to the extreme – a Californian driver pays money into his account at a drive-in bank.**

LEFT **Drivers have to pay money at toll booths in order to use some motorways.**

flowing. New, wide motorways have been built to take the extra vehicles. These motorways have been designed for safety, with two carriageways — separated by a crash barrier — to keep apart traffic moving in opposite directions. Motorways have also been designed for speed. They have gentle slopes and bends so that the traffic can keep moving, and several lanes to allow faster-moving vehicles to overtake slower traffic. Motorways are expensive to build and maintain, so in some countries drivers are charged a toll to use them.

DID YOU KNOW?
- **Vehicles in the USA use 44 per cent of the world's petrol.**
- **The narrowest road in the world is in Italy and is just 43 cm wide.**
- **The total length of roads in the USA is equal to six trips to the moon.**

At the East Los Angeles Interchange, the volume of traffic rises to 522,000 vehicles per day.

MOTOR SPORTS

Almost all vehicles are raced. Racing encourages vehicle manufacturers to experiment with new developments and test them in competition. Many of the features of modern cars, such as **disc brakes**, **fuel injection** and **independent suspension** were first tried out on racetracks, and then developed for general use.

There are basically two groups of car races. The first group, which includes rallies, is for standard factory-produced cars which have been changed for racing. The second group is for cars which are specially built for racing. This group includes many different types of racing for a huge variety of cars. These range from highly-tuned Grand Prix cars to tiny go-karts which are powered by lawn mower engines.

Car racing generally takes place on specially-built racetracks. But some races, such as the Monaco Grand Prix, are held on temporary courses around city streets. Racing cars are divided into classes so that cars with similar equipment and engines of equal size and power compete against each other.

ABOVE 'Monster truck' shows are very popular in the USA.

RIGHT 'Taking off' on a motorbike racetrack.

PROBLEMS OF ROAD TRAVEL

Over 400,000 hectares of countryside are now covered by roads. City centres are strangled by traffic during the day. Parking restrictions, such as meters and clamping, bus lanes, red routes where parking is forbidden, and computerized traffic signals have been introduced to try to ease the flow of vehicles. But these measures have had only temporary success as the number of cars increases.

More traffic has led to more accidents. Although modern cars have built-in safety features, road deaths worldwide are over 25 million per year.

The rush hour in Kyoto, Japan.

Cars pour 2,500 tonnes of pollution into the air per year in a large city like New York, creating smog above the skyline. This is an unusually clear day.

Cars cause environmental problems even when they are no longer roadworthy. Some will sit and rust in scrap heaps like this.

Motor vehicles also cause great damage to the environment. When petrol or diesel is burned it releases carbon dioxide into the atmosphere, which contributes to the greenhouse effect. The fuel used by motor vehicles causes other problems. Oil is transported around the world, to be used mainly in the tanks of automobiles. Every year about 3.5 million tonnes of this oil is spilled into the ocean, causing damage to marine life.

Some types of petrol contain lead. Burning this petrol in an engine releases the lead into the environment. The lead destroys plants and causes kidney disease, nerve disease and brain damage in humans. Road traffic also causes noise pollution.

DID YOU KNOW?
- **There are over 120 million cars in the USA – twenty-three cars per kilometre of road. In Hong Kong there is a car for every 4.1 m of road! If all the drivers drove their cars at once, the whole city would be jammed solid.**
- **The longest traffic jam in the world was 175 km. It happened in France in 1980.**
- **One of the biggest motorway crashes occurred in 1971 on the M6 in Britain. During thick fog, 200 vehicles smashed into each other. Eleven people died and another 60 were injured.**

THE FUTURE OF MOTOR TRANSPORT

Small electric-powered cars are highly suitable for short journeys round city streets – especially for parking. They could also help us to save energy.

The world's oil resources will probably run out by about the year 2030. Without petrol to power them, what kind of future will there be for motor vehicles? For the car to survive, this problem will have to be tackled. One solution would be to use the remaining resources as efficiently as possible by building much lighter, more fuel-efficient cars. They might be powered by diesel mixed with oil from sunflowers, which is already used to make a cheap fuel.

Another solution would be to use completely different fuels. **Synthetic**

DID YOU KNOW?
One system of traffic control is already operating in overcrowded Lagos, Nigeria. Only cars with even-numbered licence plates can be driven on Mondays, Wednesdays and Fridays. Cars with odd-numbered plates are allowed on the roads only on Tuesdays, Wednesdays and Thursdays.

Solar-powered cars are powered by energy from the sun.

One of the latest production cars on show at the Italian Motor Show.

fuel made from coal is already used in South Africa. An alcohol fuel called Alcool, which is made from sugar cane, or methane gas, which can be produced from rotting rubbish or sewage, are other possibilities. The solution may lie in the increased use of solar-powered and electric cars.

However, governments may have to take measures to force drivers off the road and on to mass-transport systems, such as the railways or trams. Governments in many countries are already taking action by making public transport more attractive and encouraging the transport of goods by rail. In Japan, computer systems are being used to identify individual vehicles on roads and drivers are charged a toll or a tax for using the road. It may not be long before traffic control systems like this are introduced in other parts of the world in an attempt to reduce the traffic on heavily congested roads.

GLOSSARY

Aerodynamic Having a smooth or streamlined shape, which cuts down wind resistance and makes a vehicle go faster.

Assembly line A method of production where a car passes through a factory on a slow-moving conveyor belt while workers fix parts on to it.

Disc brakes Brakes which work by a pair of pressure pads called callipers. These move together to pinch a metal disc attached to each wheel.

Fuel injection A system for squirting fuel directly into the cylinders of an engine. The fuel is pumped through nozzles, instead of through a carburettor, to create a fine spray of fuel and air in each cylinder.

Gears Toothed wheels which carry power from one part of a machine to another. Gear wheels of different sizes allow different parts of the machine to work at different speeds.

Hydraulic Worked by water or some other liquid.

Ignition An electrical system in an engine that makes a spark to burn the fuel-air mixture which goes into the cylinders.

Independent suspension A system of springing which allows all the wheels to move up and down separately as the vehicle crosses a bumpy surface. This system gives the passengers a particularly comfortable ride.

Industrial Revolution A time when machines replaced people and animals as the main source of power. It began in about 1750 and lasted for over 100 years.

Mass-production Making things in very large numbers.

Pneumatic tyres Tyres which are hollow and filled with air.

Power-steering A steering system which uses the power of the engine to help the driver turn the steering wheel.

Production car A car that is made in large numbers.

Prototype The first full-size working model of a vehicle.

Radial engine A type of engine where the cylinders are arranged in a circle, rather than a straight line as in a normal engine.

Scale models Miniature versions of cars which are in exact proportion to the full-size cars they represent.

Scanned Examined very carefully.

Speedometer A device fitted to a vehicle that measures the speed of travel.

Synthetic Man-made, not natural.

Test-bed An area used for testing machinery.

Transmission system Gears and shafts which carry the power of the engine to the driving wheels.

Turbocharger A fan, turned by the exhaust gases of an engine, which forces extra fuel and air into the cylinder to increase the engine's power.

BOOKS TO READ

Battle for the Planet by André Singer
(Pan, 1987)
Cars by Andrew Langley (Franklin Watts,
1987)
Cars and Trucks by John Fletcher
(Kingfisher, 1982)
Granada Guide to Motorcycles by Geoff
Aspel and Bill Gunston (Granada, 1981)
How Things Work by Steve Parker
(Kingfisher, 1990)
Land Transport by Malcolm Dixon
(Wayland, 1991)
Making a Motorway by James Dalaway
(Wayland, 1992)
Transport by Robin Kerrod (Wayland,
1991)
The Way Things Work by David
Macauley (Dorling Kindersley, 1988)

Picture acknowledgements

The publisher would like to thank the following for providing the pictures used in this book: All-Sport UK 6 (Gerard Planchenault); The Hutchison Library 23; Photri 20, 21 bottom (B. Howe), 24; Quadrant 12, 16 (Auto Express), 17 bottom (Auto Express), 18 (Auto Express), 19 top (Auto Express), 21 top, 25 (C. Kirkham), 28 (Auto Express), 28 middle (Auto Express); Topham Picture Library 5, 17 top (M. Mann), 24 (S. Holland); Wayland Picture Library 7 (Richard Sharpley), 9, 13, 14 (Richard Sharpley), 15, 22 bottom, 26 top (Jimmy Holmes); Zefa Picture Library COVER, 22 top, 26 bottom, 27.

All artwork by Nick Hawken.

INDEX

The numbers that appear in **bold** refer to captions.